# Contents



Which type of change - physical or chemical - occurs when wood is burned? Find out on page 5!

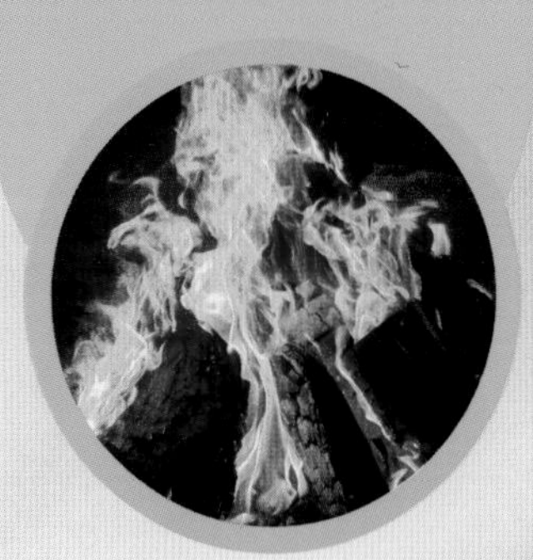

What does this car's fuel have to do with chemical reactions? Go to page 20.

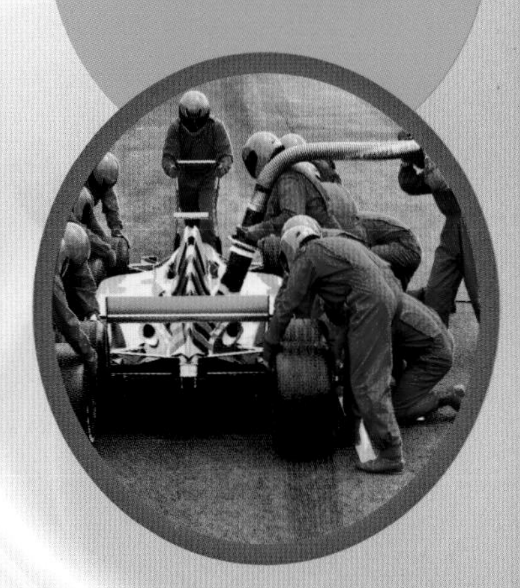

# Chemical reactions

**Matter is what makes up the world around you. And just like that world, matter is always changing! Sometimes matter changes only in shape or size. But during a chemical reaction, it changes from one type of matter to another. Chemical reactions also involve energy, often in the form of heat or light. Fireworks exploding and logs burning are both examples of chemical reactions. Other chemical reactions take place inside every living thing, including you!**

## Changing matter

How can you change a sheet of paper? You could write on it. You also could fold or tear it. These are examples of **physical changes**. The paper changes shape or size, but remains the same type of matter.

If you burn paper, however, it changes into gases and ash. This newly made matter has different properties from the original paper. For example, gas and ash cannot be folded or torn as paper can be. You cannot write or draw on gas or ash. For this reason, burning paper is an example of a **chemical change.**

When matter undergoes a chemical change, it becomes a new type of matter with different properties from those it had before. A specific example of a chemical change is called a chemical reaction.

By cutting and shaping wood, this carpenter is a master of physical changes.

# Reactions all around

Chemical reactions cook your food, clean pollution from dirty water, and even protect your teeth from cavities. In fact, while you read this sentence, a countless number of reactions are taking place inside your body to keep you alive and healthy. Chefs, dentists, doctors, and many others need to understand chemical reactions and how to control them.

**How can you tell a physical change from a chemical change? The table below lists several examples.**

| Two types of change | |
|---|---|
| **Physical change** | **Chemical change** |
| Water freezing into ice | A carbonated drink releasing bubbles |
| Mixing metal nails of different sizes and shapes in a jar | Metal nails rusting or tarnishing in the air |

Wood changes into new matter when it is burned. This is an example of chemical change.

# Explaining reactions

**Matter is made of atoms. The negatively charged particles of an atom are called electrons. Electrons are often shared or transferred among atoms. This allows atoms to join together in a chemical bond, often forming a unit called a molecule. When chemical bonds break, the molecule also breaks.**

In a **chemical reaction**, old bonds break and new bonds form. This changes one set of **compounds**, called **reactants**, into new compounds, called **products**.

Look to the right. See the bonds that break and form when sodium metal reacts with chlorine gas? Notice that chemical bonds involve the movement of electrons. The atom's central core is called the **nucleus**. It has a positive charge. The nucleus does not change when bonds break or form.

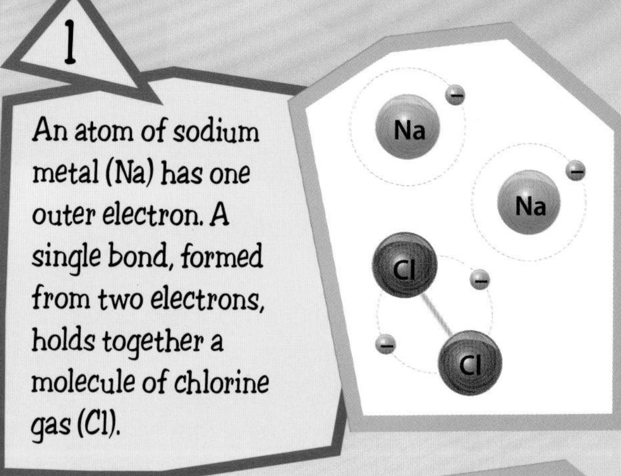


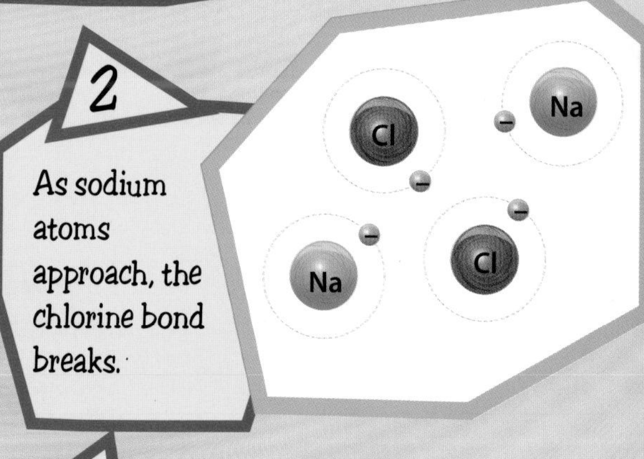


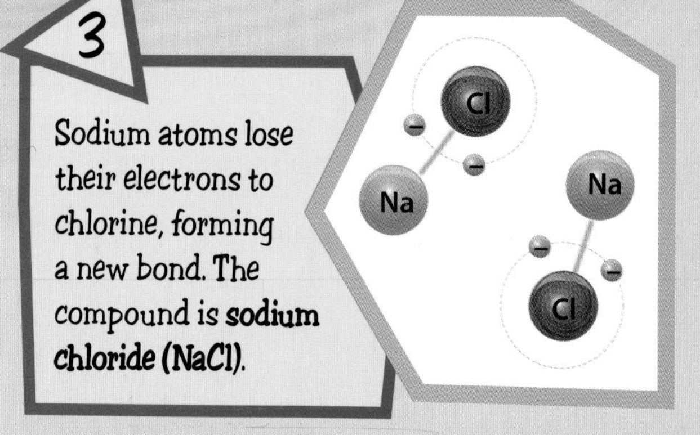

# Reactions and energy

Have you ever baked bread? You first mix flour, water, and other ingredients. This makes dough, which must be baked in an oven. The change from dough to baked bread is a chemical change. For this change to take place, the dough needs **energy** in the form of heat.

All chemical reactions involve changes in energy. Some chemical reactions release energy, while others take it up. In addition, almost every reaction needs some energy just to get started. You will learn much more about reactions and energy on page 18.

Bread rises and bakes because of chemical changes.

# Chemical reactions

To describe a chemical reaction, chemists use a type of sentence called a **chemical equation**. A chemical equation uses words or symbols to show the changes that take place during a chemical reaction.

Chemists use many different types of chemical equations. One of the simplest is called a word equation. Here is the word equation for the reaction that you studied on page 6.

sodium + chlorine → sodium chloride

The plus sign means "reacts with", and the arrow means "produces". The arrow points from left to right to show how the reaction proceeds. Sodium and chlorine are the reactants. They are the substances that begin the reaction. Sodium chloride is the product. It is the substance that the reaction produces.

# Formula equations

To save time, chemists usually write equations with chemical symbols instead of words. The symbol for each element is shown in the **periodic table** of **elements.**

An equation that uses symbols is called a **formula equation**. Here is a simple formula equation for the reaction between sodium and chlorine.

$Na + Cl \rightarrow NaCl$

This equation is read just like the word equation: sodium (Na) reacts with chlorine (Cl) to produce sodium chloride.

Open a can of fizzy drink, and the pressure inside the can is released. This allows carbonic acid ($H_2CO_3$) to break apart, forming bubbles of carbon dioxide ($CO_2$) and water ($H_2O$).

## Coefficients

**Usually, to make formula equations more useful, chemists add coefficients to balance the equation. Coefficients are numbers placed in front of the chemical symbols. They show that the same number of each type of atom will enter and leave a reaction. Look to the right. The number 2 in front of CO is a coefficient.**

Methane ($CH_4$) is a natural gas. It can be used in hobs as a fuel. Like other fuels, it reacts with oxygen to form carbon dioxide and water.

## REVERSIBLE REACTIONS

A **reversible reaction** takes place in both directions between reactants and products. To describe one, chemists use a double-headed arrow. Here is an example:

$$2CO + O_2 \rightleftharpoons 2CO_2$$

This equation shows how carbon monoxide (CO) reacts with **oxygen** ($O_2$) to form **carbon dioxide** ($CO_2$). As the double-headed arrow shows, the reverse reaction may also take place. Carbon dioxide can break apart to form the original two reactants.

# Conservation of mass

Chemical reactions form when atoms break old bonds and form new bonds. Atoms neither appear nor disappear in a reaction, they merely join together in new ways. This fact explains a principle called the **conservation of mass**. In every chemical reaction, matter is neither made nor destroyed. The **mass** or weight of the reactants equals the mass of the products.

You might wonder if this principle holds true when things burn. In a campfire, burning logs leave behind a small pile of ashes. The ashes weigh much less than the logs! However, the products of a burning reaction are more than they may appear to be. Look at this equation:

$$C_6H_{12}O_6 + 6\,O_2 \rightarrow 6\,CO_2 + 6\,H_2O$$

The cellulose in wood reacts with oxygen to form two gases: carbon dioxide and water. You see these gases as the glowing yellow flames of the fire. If you included the masses of all the gases in the reaction, you would find the conservation of mass holds true.

# Conservation of energy

Just as a chemical reaction conserves matter, it also conserves energy. In every chemical reaction, energy is neither made nor destroyed. However, energy almost always changes form in a reaction.

Look again at the reaction of burning logs on the previous page. This reaction releases a lot of energy in the form of heat and light. Even more energy is released when hydrogen burns in rockets, such as those that launch a space shuttle. Hydrogen is the fuel used in a space shuttle. The energy is enough to send it into space!

Where does all this energy come from? The answer is the chemical bonds inside the fuel. The bond inside hydrogen gas ($H_2$) holds a large amount of energy. The rocket combines this energy with oxygen gas ($O_2$). When these bonds break, that energy is released. Look at the equation in the picture below. If you added up the energy that resulted, you will also find that the **conservation of energy** holds true!

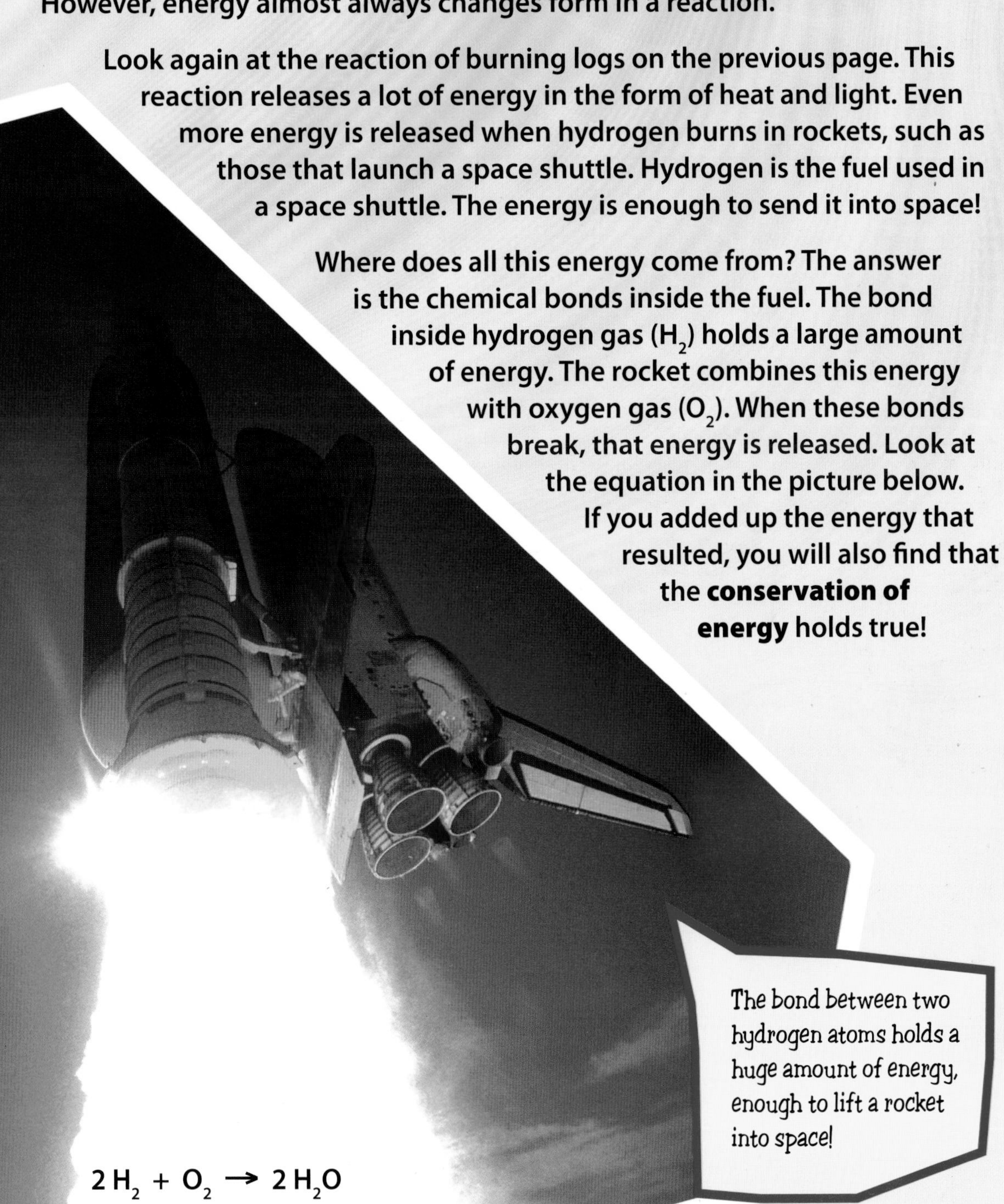


The bond between two hydrogen atoms holds a huge amount of energy, enough to lift a rocket into space!

# Chemicals of life

Some chemical reactions, such as the burning of wood, are easy to observe and study. Other reactions are hidden. When you eat an apple, for example, you cannot see it changing inside your body.

In the early 1800s, chemists believed that the reactions inside living things were very different from other reactions. They believed that living things held some special force. The force allowed the bodies of plants and animals to make compounds that could not be made elsewhere. Compounds are formed when at least two different **elements** are joined in a chemical reaction.

In 1828 German chemist Friedrich Wohler showed that this idea was wrong. By combining chemicals in his laboratory, Wohler made a compound called urea. Urea is one of the compounds found in living things. Wohler used no "special force" to make it.

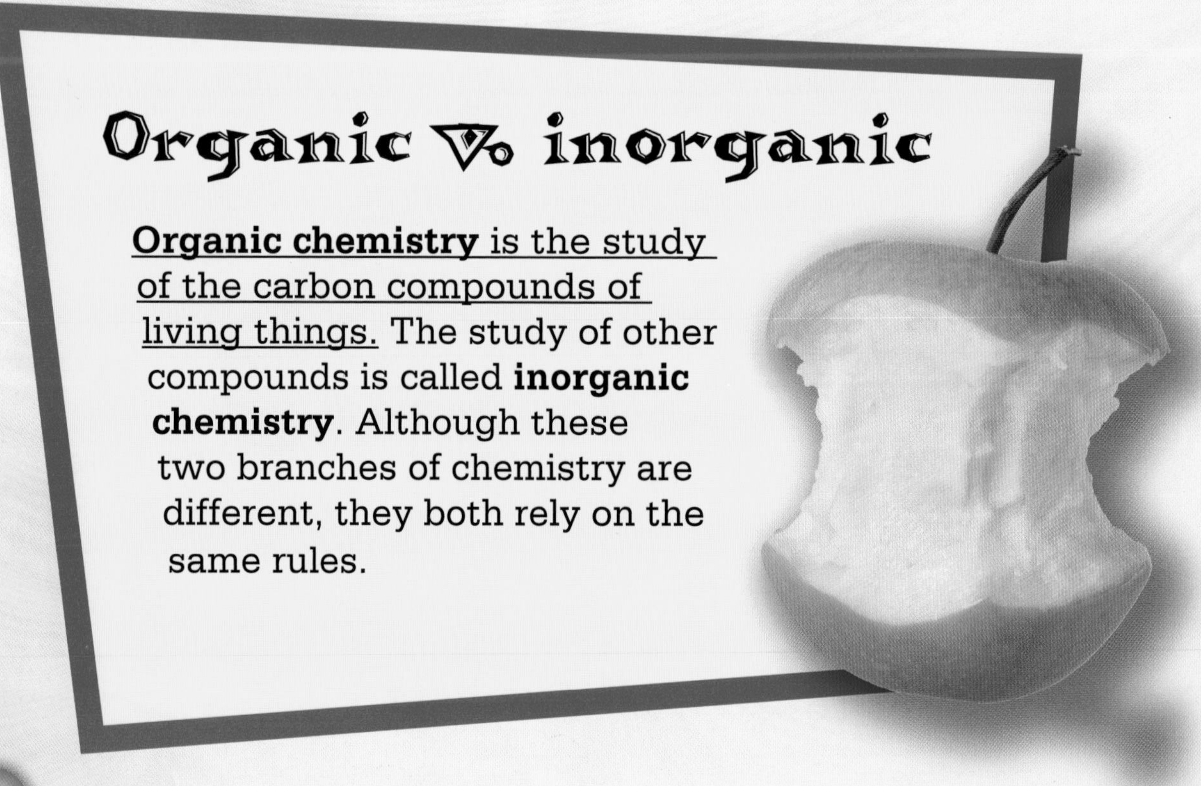

## Organic vs inorganic

**Organic chemistry** is the study of the carbon compounds of living things. The study of other compounds is called **inorganic chemistry**. Although these two branches of chemistry are different, they both rely on the same rules.

# Carbon compounds

Today, chemists know a lot about the reactions that take place in living things. They have learned that living things rely on an important element called carbon. Living things make and use a huge number of compounds that contain carbon.

Why is carbon important? Unlike the atoms of other elements, carbon atoms can form long chains. These chains form the wide variety of molecules that living things need and use.

Safety always comes first when chemists run experiments. This chemist wears safety goggles, a lab coat, and protective gloves. She also keeps her long hair pulled back.

# Classifying reactions

## Synthesis and decomposition

In a **synthesis reaction**, two or more **reactants** combine to form a single **product**. The artwork below shows the reaction of iron (Fe) and sulphur (S) to form iron sulphide (FeS). The reaction of sodium and chlorine shown on page 6 is also a synthesis reaction.

**Synthesis:** $Fe + S \rightarrow FeS$

A synthesis reaction run backwards is called a **decomposition reaction.** In a decomposition reaction, a single reactant breaks apart into two or more products. One example is the decomposition of calcium carbonate ($CaCO_3$) into two **compounds**.

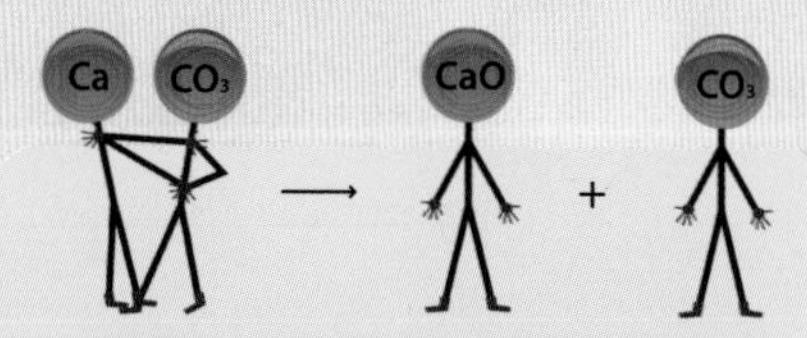


**Decomposition:** $CaCO_3 \rightarrow CaO + CO_2$

You can compare a synthesis reaction to two dancers forming a pair on the dance floor. In a decomposition reaction, the dancers say goodbye and go their separate ways.

Tri-nitro toluene (TNT) breaks apart in a decomposition reaction. It can release enough energy to bring down a building!

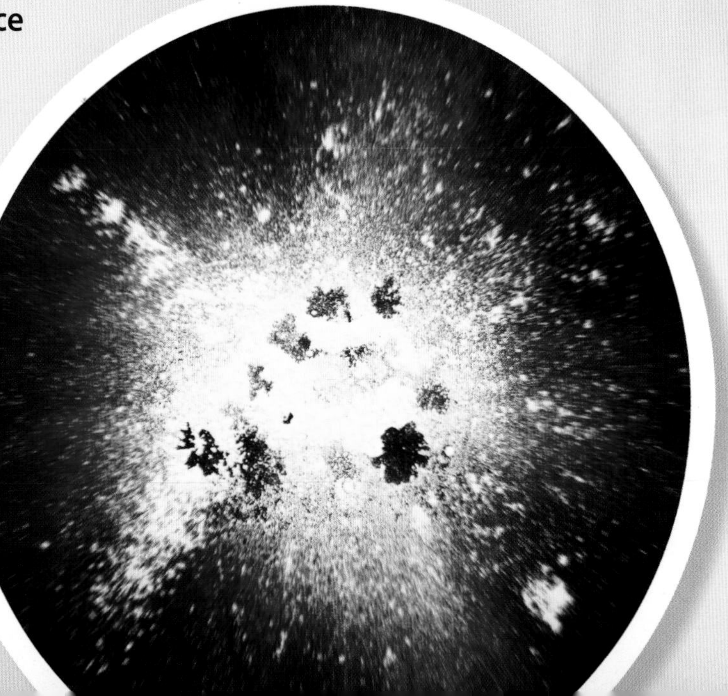

# Replacement reactions

There are two types of replacement reactions. In a **single replacement reaction**, one reactant takes the place of another in a compound. Below, magnesium (Mg) replaces copper (Cu) in a compound with a sulphate **ion** ($SO_4$).

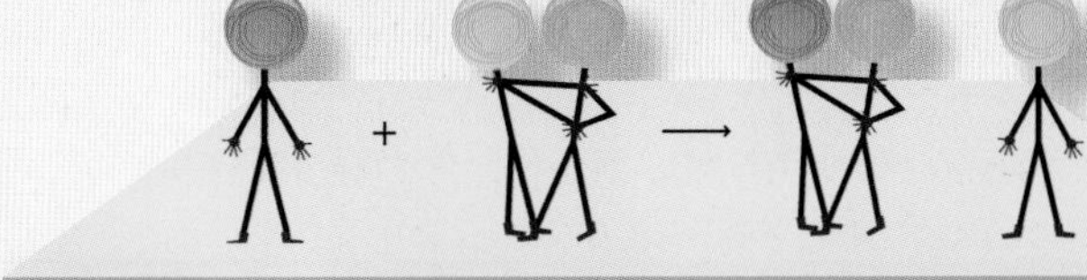

**Single replacement:** $Mg + CuSO_4 \rightarrow Cu + MgSO_4$

In a **double replacement reaction**, two reactants replace each other in two compounds. The artwork on the right shows silver (Ag) and potassium (K) replacing one another.

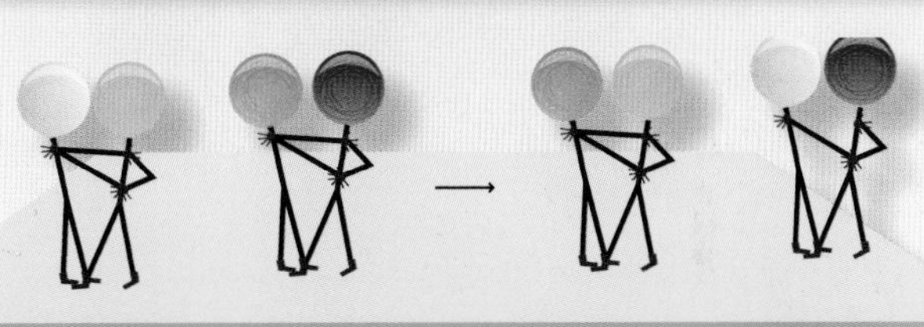

**Double replacement:** $AgNO_3 + KCl \rightarrow KNO_3 + AgCl$

You can compare both of these reactions to dancers changing partners. A single replacement involves three dancers, while a double replacement involves four.

## Reactivity series

In the single replacement reaction in the artwork, the reaction goes from left to right, rather than right to left. This happens because magnesium is more chemically reactive than copper. In most single replacement reactions, the more reactive **element** replaces the less reactive element.

The list on the right is called a reactivity series for the metallic elements. The metals are listed from the most reactive (lithium) to the least reactive (gold).

**MOST REACTIVE**

Lithium (Li)
Potassium (K)
Barium (Ba)
Strontium (Sr)
Calcium (Ca)
Sodium (Na)
Magnesium (Mg)
Aluminium (Al)
Manganese (Mn)
Zinc (Zn)
Iron (Fe)
Cadmium (Cd)
Cobalt (Co)
Nickel (Ni)
Tin (Sn)
Lead (Pb)
Copper (Cu)
Silver (Ag)
Mercury (Hg)
Gold (Au)

**LEAST REACTIVE**

# Combustion reactions

Not all **chemical reactions** fit into the four categories that you just learned. An example is any reaction that involves burning, which is called a **combustion reaction.** In a combustion reaction, a fuel releases heat as it combines with **oxygen**. The typical products of the reaction are water and **carbon dioxide**.

The burning of wood is an example of a combustion reaction. The burning of hydrogen gas, which is used as rocket fuel, is another. Burning hydrogen is unusual because it does not produce carbon dioxide.

Other combustion reactions take place gradually. For example, after you swallow food, your body uses a long series of reactions to break it apart. The reactions slowly release the food's energy, much of which your body stores. Or, if you are an athlete, the food fuels your body!

The food you eat combines with the oxygen you breathe in. The products include carbon dioxide, which you breathe out.

Food + $(O_2) \rightarrow (CO_2) + (H_2O)$

## Oxidation-reduction reactions

An **oxidation-reduction reaction** is often called a **redox reaction**, for short. In a redox reaction, **electrons** move from one compound or element to another. You already have seen examples of this reaction. Many synthesis, decomposition, and single replacement reactions are redox reactions.

Certain redox reactions are used to make a device called an **electrochemical cell**. These cells can be joined to form **batteries**, such as the batteries you use in toys and radios. Electrons are released in a reaction on one side of the cell. Then they move across a wire to a reaction on the other side. That movement of electrons is electricity!

## Acid-base reactions

**Acids** and **bases** are two classes of compounds. In many ways, their properties are opposite from one another. Acids, such as orange juice or vinegar, taste tart. Bases, such as baking soda or soapy water, taste bitter.
**SAFETY NOTE:** Never taste or swallow laboratory chemicals!

When an acid reacts with a base, the product has properties very different from either an acid or a base. An acid and a base will react to neutralize each other's properties. Neutralize means to reduce or eliminate. You will lean more about acids, bases, and their reactions on page 30.

# ENERGY and CHEMICAL REACTIONS

**All chemical reactions involve chemical bonds breaking and forming. These changes involve energy, which is why chemical reactions either release energy or take it up. Typically, this energy flows in the form of heat. Heat is the energy that is moved from one object to another because of a difference in temperature.**

## Exothermic reactions

Many reactions release heat, which means they make their surroundings hotter. Reactions that release heat are called **exothermic** reactions. The prefix *exo* means outer, and *thermic* means related to heat.

A **combustion reaction** is exothermic, as you can tell when you sit next to a fire. Other combustion reactions warm you from the inside. Your body is constantly breaking down food, both the food you eat and stored food. These reactions are exothermic, so they release heat.

## Endothermic reactions

Other reactions take up heat from their surroundings. Reactions that take up heat are **endothermic**. The prefix *endo* means inner.

Have you ever used a chemical cold pack? A cold pack can help keep down the swelling on an injured arm or leg. It works because the chemicals inside it combine in an endothermic reaction. The reaction takes up heat, so the pack feels cold.

# Units of energy

To measure the energy change of chemical reactions, scientists typically use either the unit of the **joule** or the **calorie**. The unit of the Calorie – with a capital C – is used in nutrition. A Calorie equals 1,000 calories, or 1 kilocalorie. The prefix *kilo* means thousand.

Measuring energy is important in diet and nutrition because all foods provide your body with energy. For example, as the table shows, one banana provides 105 calories. This energy changes form as your body breaks down the banana. Some of the energy is released as heat, and some is stored for future use. Your body might use the energy from one banana to run or repair itself.

## Bioluminescence

Some chemical reactions release energy in unusual ways. For example, inside a firefly, a reaction releases energy in the form of light. This is called **bioluminescence**. Unlike combustion reactions, the reactions of bioluminescence do not release much heat. They produce a "cold light" that does not burn the firefly!

| Food | Calories |
|---|---|
| Bagel | 216 |
| Medium banana | 95 |
| Pork sausage (24 grams) | 73 |
| Medium baked potato | 218 |
| Peanut butter (1 tbsp) | 95 |

Foods provide energy, as well as vitamins and minerals for your body.

# Fuels and energy

A **fuel** is a material that releases energy when its chemical or physical structure is changed. Burning is an example of a fuel releasing its energy chemically. Coal is burned at power plants to make electricity. Petrol is a fuel for cars. All fuels hold energy in their chemical bonds. The energy source for most fuels is the sun.

Most fuels began as plants. In a process called **photosynthesis**, plants use the energy of sunlight to make their own food. That food is also used to make every part of a plant, including the wood of trees.

# Fossil fuels

Coal, oil, and natural gases (such as **methane**) are examples of **fossil fuels.** Fossil fuels formed from plants that lived millions of years ago. Back then, Earth was covered in swamps and was thick with plants. When the plants died, they sank to the bottom of a swamp and were quickly buried. Over time, the plants were slowly chemically changed. Deep underground, they became fuels of much higher energy than ordinary plant **matter**.

Every year, people dig up more and more fossil fuels. This supply is turned into petrol for cars, diesel for lorries, and jet fuel.

Racing cars need a lot of petrol to move fast. Why do you think the crew is dressed in fireproof suits?

# Ethanol

One problem with fossil fuels is that they are non-renewable. This means that Earth has a limited supply of them. Once the supply is used up, it will be gone forever. Unfortunately, the world is burning fossil fuels faster than ever before!

To help conserve fossil fuels, people are making and using a **compound** called **ethanol** ($C_2H_6O$). Ethanol is a useful fuel. Although it is not as rich in energy as petrol, it can be made from ordinary plants, such as corn. Scientists continue to study ways to make and use ethanol as efficiently as possible.

This ethanol plant changes corn kernels into ethanol. Most petrol sold today has at least some ethanol mixed into it.

# Too much burning?

For most of human history, people mainly burned wood to meet energy needs. But in the 1800s, coal became a popular fuel for ovens and furnaces. Later, engines that ran on petroleum products (such as petrol) became widely used. Today, people are burning more fossil fuels than ever before.

What's the harm of burning fossil fuels? Remember that combustion reactions produce a gas called **carbon dioxide ($CO_2$)**. In small amounts, carbon dioxide is a natural part of Earth's atmosphere. It fills an important role, too, by helping trap Earth's heat in the atmosphere. This is called the **greenhouse effect**.

Unfortunately, by burning large amounts of fossil fuel, people are adding more and more carbon dioxide to Earth's atmosphere. Scientists say that the higher levels of carbon dioxide are increasing the greenhouse effect. This is causing **global warming**.

For the past 150 years, burning fossil fuels has been dumping more and more carbon dioxide ($CO_2$) into Earth's atmosphere.

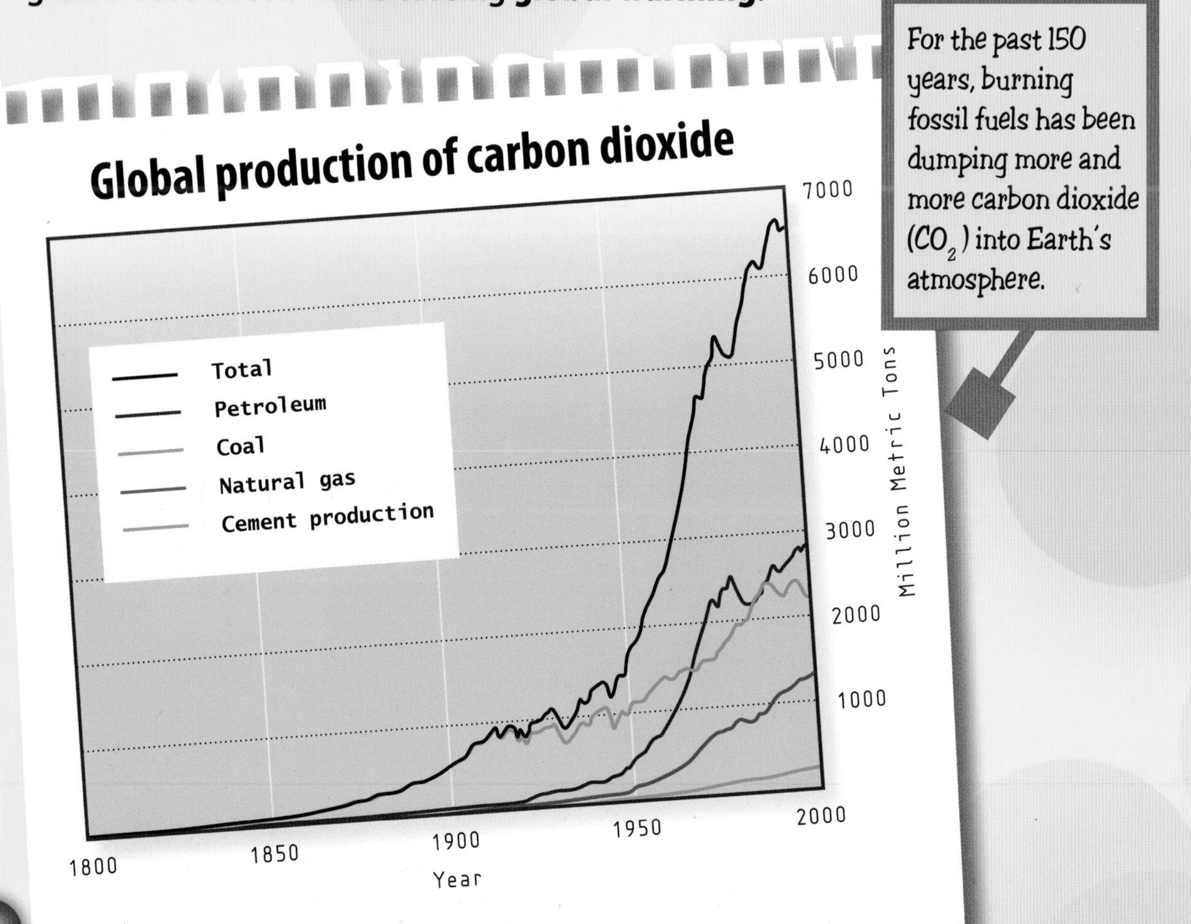

# Fighting global warming

Global warming is a dangerous problem. Weather data shows that Earth's winter weather is warming and summer weather is becoming even hotter. Ice caps and glaciers are melting. Climate is slowly changing everywhere.

One way to slow global warming is to replace fossil fuels with other energy sources. These sources include solar power, wind power, nuclear power, and perhaps the energy from ocean tides. Geothermal energy is the energy from Earth's hot interior. It can be tapped to heat homes and businesses. For vehicles, some new models now run on petrol and electricity together. Cars that run totally on electricity are also available. These cars may become very common in the future.

Today, scientists all over the world are working on new technology and new ideas to slow global warming. You will read about one of these ideas on the next page. You will also learn about ways that you can help solve the problem.

The effects of global warming are especially strong in places near the poles.

# Methane

Carbon dioxide ($CO_2$) is not the only gas that is raising Earth's temperatures. Natural gas, or **methane** ($CH_4$), also traps heat. In fact, a **molecule** of $CH_4$ traps 20 percent more heat than a molecule of $CO_2$. Although methane is rare in the atmosphere, scientists estimate that its levels have risen 145 percent over the past 100 years.

What caused this rise? Human actions seem to be the source. On farms, cows and sheep produce methane as they digest food. Bacteria that live in rice paddies release methane as do bacteria that live in termites. When rainforests are cut down, termites feast on the stumps.

## Bicarbonate of soda to the rescue?

Some combustion reactions are the likely cause of global warming. Could other reactions be part of the solution? Scientist Joe Jones from Texas, USA, thinks they can. He invented a process to react carbon dioxide gas with a chemical called sodium hydroxide. The **product** is ordinary bicarbonate of soda, just like the type that cooks use.

Jones founded a company called Skyonic to continue developing his ideas. The company's goal is to capture carbon dioxide from coal-burning power plants. The new waste product - bicarbonate of soda - could be sold and used. It could also be buried safely in the ground.

A **synthesis reaction** changes carbon dioxide ($CO_2$) into bicarbonate of soda ($NaHCO_3$). If the reaction can be run on a large scale, it could be used to slow global warming.

# Taking action

Many scientists, government leaders, and business leaders are calling for a worldwide effort to stop global warming. The problem is global, so the solution must also be global.

What can you do? You cannot stop global warming by yourself, but you can be part of the fight. The following five ways can help to slow the release of carbon dioxide into Earth's atmosphere. Share the list with your friends, family, and classmates.

- **Conserve fossil fuels!** Cycle or walk instead of using a car, or use public transport.
- **Conserve electricity!** Switch to energy-saving fluorescent light bulbs, and turn off electric appliances when they are not in use.
- **Plant trees!** All plants take in carbon dioxide from air. Trees use it to make wood.
- **Buy locally!** Lorries and ships burn a lot of fuel to bring you products from far away. Buy locally made products instead.

## GLOBAL WARMING FACTS

- Since 1850, average temperatures worldwide have increased 0.8° C (1.4° F).
- The years from 1996 to 2007 were the warmest span of 12 years ever recorded.
- People of the United States make up 5 percent of the world's population, but drive almost 33 percent of the world's vehicles.
- Near the North Pole, sea ice has been decreasing about 3 percent every 10 years.
- In 2002 an ice shelf with a surface area of 3,250 sq km broke off from Antarctica.

This hydrogen bus produces no carbon dioxide, just water.

# CONTROLLING REACTIONS

**All sorts of chemical reactions take place in the world around you. Some of these reactions are unwanted, such as the reaction that rusted the bridge shown in the picture. Other reactions are very useful, but only at the right times or in the right order. Read on to learn three ways that reactions can be controlled.**

## Keep reactants apart

Steel is a hard, sturdy material made from iron (Fe) and other chemicals. Unfortunately, in wet, humid conditions, iron reacts with **oxygen** to form iron oxide ($Fe_2O_3$), also known as rust. Rust is soft, crumbly, and ugly. It can ruin anything made of steel.

A simple way to stop rusting is to keep the **reactants** apart. A reaction takes place only if the reactants meet. This is why people paint steel bridges and cars. The paint keeps steel away from oxygen and air.

A fresh coat of paint would have prevented the rusting reaction on this bridge.

Food spoils because of bacteria, fungi, and other tiny creatures. At low temperatures, these creatures grow slowly, if at all.

## Change the temperature

Raw meat may not be healthy to eat. But at high temperatures, chemical reactions change raw meat to cooked meat. Even if cooked meat is allowed to cool, it looks and tastes quite different from before. Many reactions take place only at certain temperatures. Cooking food, for example, requires very high temperatures. So do **combustion reactions**. This is one reason why water can put out a fire. The water lowers the temperature and stops the combustion.

Temperature changes may also speed up or slow down a reaction. The reactions that spoil food, for example, are slower at low temperatures than high ones. This is why foods keep well in a refrigerator and even better in a freezer.

### CHANGE THE SURFACE AREA

The next time you sit at a campfire, compare how different pieces burn. You will find that thin branches and twigs burn faster than big, thick logs. Why? The reason is that smaller pieces have greater surface area for their size. Surface area is where the wood meets the air. By increasing the surface area between reactants, a reaction speeds up.

# Activation energy

Why does wood burn only at high temperatures? If you toss even a thin splinter into the air, why will it not react with the oxygen it meets? The reason involves a concept called **activation energy**. Most reactions need an input of **energy**, called the activation energy, just to begin.

You can compare a reaction to a ball rolling over an "energy hill". In the diagram on page 29, the ball is like the reactants. The height of the hill is the activation energy. For the reaction to move forward, the ball must be pushed over the hill. In the case of burning wood, the "push" could come from a lit match.

The high activation energy of burning wood means that trees do not normally burn up. But if one tree catches fire, it may provide the energy for lighting the tree next to it.

# Catalysts and enzymes

Like wood, sugar will react with oxygen at high temperatures. But when you eat sugar, it reacts with oxygen inside your body, where the temperature is not that hot! How is your body able to overcome the activation energy for burning sugar? The answer involves **catalysts**.

A catalyst speeds up a chemical reaction by lowering its activation energy. A catalyst is neither a reactant nor a **product** of a reaction. Instead, it merely helps the reaction take place faster.

Examples of catalysts are all around you. **Salt**, for example, acts as a catalyst for the rusting of iron. In your digestive tract, catalysts work to break down sugar and the other foods you eat.

Catalysts that living things make and use are called **enzymes**. Nearly all the reactions of living things rely on enzymes.

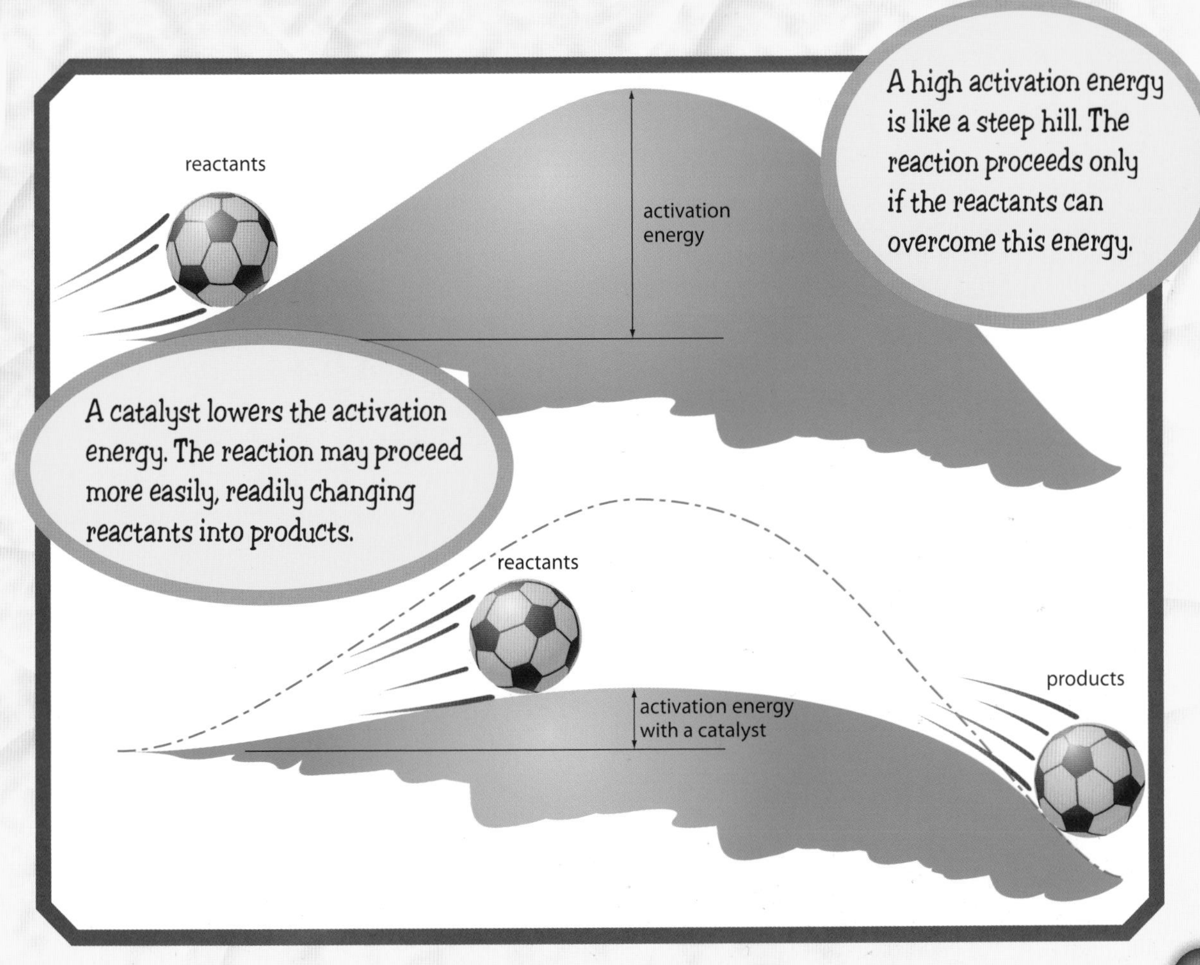

# Acids and bases

**Particles with electric charges are called ions. The simplest ion is the hydrogen ion (H+). Acids are compounds that give away hydrogen ions, while bases receive them. Some acids and bases are very strong. They must be handled carefully. Others are mild enough to be part of your food. Reactions between acids and bases take place all around you. They even happen inside your body.**

## Ions in water

Did you know that water can act as an acid or a base? In a reaction between two water **molecules**, a hydrogen ion ($H^+$) may transfer from one molecule to the other.

$$2H_2O \rightleftharpoons H_3O^+ + OH^-$$

As the double arrows show, this reaction is reversible. In fact, the two **products** change back into water very quickly. In any sample of water, the reaction is always taking place. Every sample of pure water holds small but equal numbers of **hydronium ions ($H_3O^+$)** and **hydroxide ions ($OH^-$)**. As you will see, these ions are key to understanding acids and bases.

# Reactions with acids and bases

The reaction shown on page 30 is an example of an **acid-base reaction.** In an acid-base reaction, a hydrogen ion ($H^+$) transfers from one compound to another. The acid is the compound that loses a hydrogen ion. The base gains it.

Mix an acid or base in water, and it will disrupt the balance between **hydronium** and **hydroxide ions.** The solution becomes either acidic or basic. For example, consider how hydrochloric acid reacts with water:

$$HCl + H_2O \rightarrow Cl^- + H_3O^+$$

Here, hydrogen ions move from hydrochloric acid ($HCl$) to water ($H_2O$). This forms hydronium ions ($H_3O^+$). The hydronium ions make the solution acidic.

Now study the reaction between ammonia ($NH_3$) and water.

$$NH_3 + H_2O \rightarrow NH_4^+ + OH^-$$

Ammonia acts as a base because it gains hydrogen ions from water. The water changes into hydroxide ions ($OH^-$). This makes the solution basic. Another name for basic is alkaline.

Like other strong acids, hydrochloric acid ($HCl$) dissolves in water to form hydronium ions ($H_3O^+$).

# The pH scale

To measure the strength of an acid or a base, scientists use the **pH scale**. The two letters stand for potential of hydrogen. For most compounds and solutions, pH values range from 0 to 14. Strong acids have low pH, while strong bases have high pH. Water has a pH of 7, which is the exact middle of the scale.

# Salts

In everyday language, the word **salt** means a specific compound called **sodium chloride (NaCl)**. This is the salt you put on food. But in chemistry, the word salt applies to a large class of compounds.

**When a strong acid mixes with a strong base, they neutralize each other's properties and form a salt.** For one example, study the reaction between sulphuric acid ($H_2SO_4$), a strong acid, and sodium hydroxide (NaOH), a strong base.

$$H_2SO_4 + 2\,NaOH \rightarrow Na_2SO_4 + 2\,H_2O$$

Sulphuric acid is so strong it can burn through metal, while sodium hydroxide is used as bleach. Never put either of these compounds on bare skin! But when they react, the products are much less dangerous. The products are water and sodium sulphate ($Na_2SO_4$), a salt.

## pH and you

Your body also uses acids and bases. The stomach's acid is very strong, to break down food and release nutrients. Outside of the stomach, however, acids and bases are quite mild. This is important because strong acids or bases could damage tissues easily.

The pH of your blood is about 7.4. It stays at that level because of compounds called **buffers**. A buffer reacts with acids and bases to keep pH levels steady. Should your blood pH ever rise or drop even a little, you would probably become very ill.

Vinegar is an acid. Bicarbonate of soda is a base. The bubbles show that a chemical reaction is taking place.

## Mix an acid and a base

### Wear goggles during this activity.

1. Obtain small samples of vinegar and bicarbonate of soda. Compare their appearance, feel, and other properties that you can observe. Do not mix them.
2. Add a few drops of vinegar to the bicarbonate of soda. Observe the results.
3. Describe the reaction between vinegar and bicarbonate of soda. What does this tell you about an acid-base reaction?

**SAFETY!**
**Do not substitute stronger acids or bases for the chemicals in this activity.**

# A world of reactions

**Living things use a countless number of reactions to stay alive and to grow. In addition, many use reactions to make some unusual, amazing products. Examples include rubber from rubber trees, honey from bees, and the stinky spray from skunks.**

## Photosynthesis and respiration

An acorn is smaller than the palm of your hand. How can it grow into a tall oak tree? The answer involves a process called **photosynthesis**.

Photosynthesis involves a long series of separate **chemical reactions**. When those reactions are put together, however, the overall reaction looks like this:

$6\ CO_2 + 6\ H_2O + \text{energy} \rightarrow C_6H_{12}O_6 + 6\ O_2$
(**carbon dioxide** + water + **energy** → sugars + **oxygen**)

In photosynthesis, plants join carbon dioxide and water to form sugars ($C_6H_{12}O_6$) and oxygen. The energy for this process comes from sunlight. Plants use the sugars for food.

Plants also provide food for people and other animals. In a process called **respiration**, both plants and animals break down food for energy. Like photosynthesis, respiration also takes place in a long series of reactions:

$C_6H_{12}O_6 + 6\ O_2 \rightarrow 6\ CO_2 + 6\ H_2O + \text{energy}$

Did you notice anything unusual about the reactions for photosynthesis and respiration? They are the opposites of each other! This is one very important way that plants and animals depend on one another.

# Reactions and you

Living things are made of tiny units called cells. Your body has trillions of cells. Each cell does a special job to keep you alive and healthy. In the liver, cells use reactions to rid the blood of wastes and poisons. When it's time to grow, a countless number of reactions make new cells from old ones.

Chemical reactions take place both inside your body and in the world around you. Learning how reactions work is an important key to understanding life on Earth.

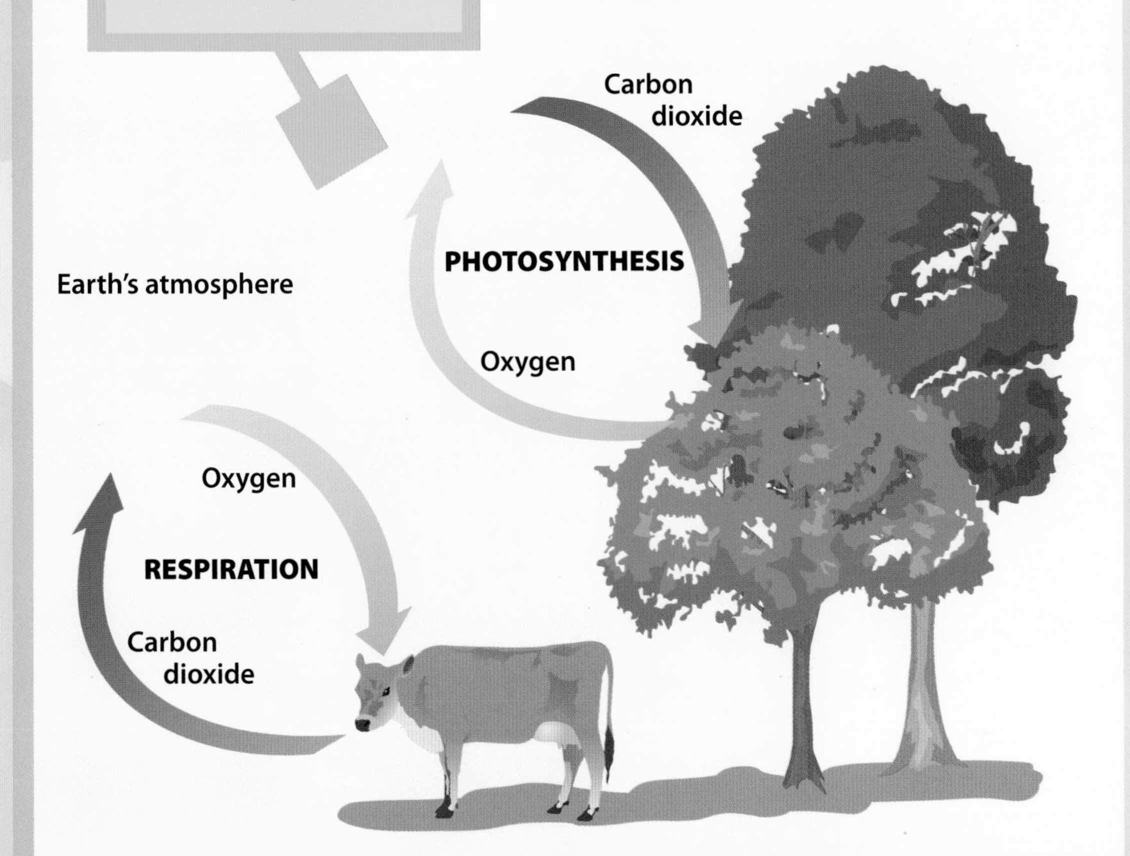


Because of photosynthesis and respiration, carbon dioxide and oxygen cycle among plants, animals, and Earth's atmosphere.

# Reactions in the air

Every day, gases are released from Earth's surface and rise into the air. Water vapour evaporates from oceans and lakes. Plants add oxygen to the air, while animals add carbon dioxide. Burning **fossil fuels** adds carbon dioxide, too. All of these gases are natural parts of the atmosphere.

Yet human actions have been releasing other gases. Factories, vehicles, and coal-burning power plants release nitrogen and sulphur **compounds** into the air. When these compounds mix with water, their reaction has a dangerous effect.

For example, sulphur trioxide ($SO_3$), a sulphur compound, reacts with water in this reaction:

$$SO_3 + H_2O \rightarrow H_2SO_4$$

The product is sulphuric acid ($H_2SO_4$), one of the strongest of all **acids**. A similar reaction produces nitric acid ($HNO_3$), another strong acid. Acids of sulphur and nitrogen mix with water in the air, then fall to the surface as **acid rain.** Acid rain has caused much damage to forests and rivers. It has also damaged statues and buildings.

Years of acid rain have eaten away at this statue, causing damage.

# Reactions underground

What types of materials lie deep underground? Diamond and sapphire are beautiful gems. Marble and granite are hard rocks used to make floors and countertops. And as you have read, coal, oil, and natural gas are valuable **fuels**.

Each of these materials formed, at least in part, from chemical reactions that took place below Earth's surface. Chemical reactions change matter underground, just as they do elsewhere. Gems and fossil fuels can form underground because conditions are very different from those on the surface. Temperature and pressure are very high, and there is no air. In addition, reactions underground often proceed very slowly. A rock may take up to millions of years to form!

## Algal blooms and the Dead Sea

Algae look like tiny floating plants. Most lake and ocean waters hold at least a small number of algae. But sometimes, compounds of nitrogen or phosphorus increase rapidly in the water. When this happens, the number of algae may grow very quickly. This event is called an **algal bloom**. An algal bloom kills plants and fish by blocking sunlight and depleting the water of oxygen. The algae may also release a dangerous toxin.

An algal bloom shows how water quality can affect living things. Life in water will thrive only under certain conditions. In the Middle East, the Dead Sea is too salty to support life. It is home only to certain bacteria and fungi.

The Dead Sea cannot support plant and animal life.

# Reactions to make new products

Many chemical reactions happen in nature. However, others do not take place until people invent them. New reactions might be used to invent a medicine or a new type of plastic. Or they can be used to find a new, better way to make a product. Here are three products that depend on reactions that people invented:

## Fertilizers

In the years before World War I, German scientist Fritz Haber was studying ammonia ($NH_3$). Germany needed ammonia to make explosives. Haber discovered a process to raise the amount of ammonia that the chemical reaction produced.

Today, the **Haber Process** is used to make over 100 million tons of ammonia and related compounds every year. But these compounds are not used for explosives. Instead, they are put into fertilizer and plant food! Plants need these compounds to grow. By spreading fertilizer, farmers and gardeners help crops grow and produce more food.

## Batteries

**Batteries** power toys, torches, and computers. A car uses a large battery to start its engine. Some cars run on both batteries and petrol engines. In the future, cars that run only on batteries may become more practical. Batteries use oxidation-reduction, or redox reactions, to make electricity. In torch batteries, one compartment holds powdered zinc (Zn). A separate compartment holds manganese dioxide ($MnO_2$). When the battery is connected into an electric circuit, zinc releases **electrons** and manganese dioxide gains them. The moving electrons make electricity.

Other batteries are made with metals such as lead (Pb) and nickel (Ni). Scientists continue to look for ways to make batteries more powerful and more useful.

# Plastics

**Plastics** can be as hard and tough as a bowling ball, or as thin and flexible as a bin bag. Some plastics make clearer, stronger windows than a pane of glass. Other plastics are useful fabrics for shoes or raincoats.

What is a plastic? Plastic is the general term for a class of human-made **polymers**. A polymer is a huge **molecule** made from small units that repeat over and over again. In recent years, scientists have developed many new polymers and uses for them. But nature has been making polymers for much longer. **Proteins** and other vital compounds are examples of the polymers in living things.

Plastics can be recycled. This means they can be used over and over again.

# REACTIONS for you to try

**You do not need a fancy laboratory to study chemical reactions. You can run many reactions in your own kitchen or garden! Try the reactions on these two pages. However, remember to follow all safety rules. These rules include following directions and using materials exactly as they are written.**

## Raisin races

What happens to a raisin in a glass of fizzy drink? Follow these directions to find out.

**MATERIALS**
a bottle of clear fizzy drink, 3 raisins, 3 drinking glasses or other containers of different sizes, timer

1. Open the bottle and fill a container with fizzy drink.
2. Drop a raisin into the container. Put the cap back on the bottle. Count how many times the raisin rises and falls in 5 minutes.
3. Do you think the size of the container changes how long a raisin keeps moving in fizzy drink? Test to find out.

As this experiment shows, raisins are just light enough to hitch a ride on bubbles of **carbon dioxide**. The bubbles form from the decomposition of carbonic acid, the **compound** that gives fizzy drink its tart taste. The reaction begins as soon as the bottle is uncapped.

# Protect clippings

Have you ever clipped and saved an article from the newspaper? Over time, the clipping changes. First it becomes yellow. This happens because of lignin, which is a compound naturally found in wood. But the paper also becomes more brittle. In fact, after many years, a newspaper clipping may break apart in your hands!

Newspapers slowly break apart because of aluminium **salts** in the paper. Paper makers add salts to help keep ink from blurring. But over time, they pick up water from the air and form **acids**. The acids slowly eat away at the paper.

For a document to last a long time, choose acid-free paper. You also can treat newspaper clippings with a basic solution that neutralizes acids. Follow the directions below.

**SAFETY**
**Perform this activity only with the permission of your parent, teacher, or guardian.**

Materials:
Newspaper, milk of magnesia, soda water, measuring spoon, large pan, towel

1 Mix 1 tablespoon of milk of magnesia into a bottle of soda water. Let it stand overnight.

2 Pour enough of the mixture to cover the bottom of the pan. Place newspaper clippings in the pan and let them soak for 1 hour.

3 Carefully pick up each clipping and let the water drip off. Dry them on a paper towel.

4 After drying, save the clippings in your scrapbook. They will last a long time!

# Secret writing

You can use lemon juice to write a secret message! This is because lemon juice weakens paper, but it does not leave a coloured stain. Heat causes the weakened part of the paper to burn slightly, colouring your message brown.

Materials: lemon juice, white paper, toothpick, lamp

1. Dip the toothpick in lemon juice. Use it to write a message on the paper. Keep dipping the toothpick to keep lemon juice on it.
2. Let the paper dry. To reveal the secret message, warm it under a lamp or in bright sunlight.

**SAFETY**
Do NOT touch paper to a hot light bulb.

# Rusting rates

Materials: Iron nails, water, salt, 2 glasses, 1 plate

1. Find three identical iron nails. Draw pictures of them.
2. Place one nail in a glass of water. Place the second nail in a glass of salt water. Keep the third nail on a plate, away from water.
3. Observe the nails over the course of a day, a week, or longer. Draw pictures and record your observations.
4. How do you explain the differences you observed between the three nails?

**SAFETY**
Do NOT touch the nails during the experiment.

# Chemical reactions review

- Chemical reactions change **matter** from one form to another.
- Both matter and **energy** are conserved during a chemical reaction. They change form, but are neither created nor lost.
- Burning is an example of a **combustion reaction**. A fuel is combined with **oxygen** to release **carbon dioxide** ($CO_2$) and water ($H_2O$).
- Reactions that release heat are called **exothermic**. Reactions that take up heat are called **endothermic**.
- Reactions may be controlled by separating **reactants**, changing the temperature, and changing the surface area of the reactants.
- Most chemical reactions proceed only with an input of energy, called the **activation energy**. A **catalyst** acts to lower the activation energy, allowing reactions to take place readily.
- In an **acid-base reaction**, a hydrogen **ion** ($H^+$) transfers from one compound, called an acid, to another compound, called a **base**.
- Chemical reactions take place nearly everywhere: in air, in water, underground, and inside the human body.

# Glossary

**acid** compound that releases a hydrogen ion in a chemical reaction

**acid-base reaction** reaction in which an acid and a base react together, neutralizing each other's properties

**acid rain** rain, snow, or other precipitation mixed with sulphuric acid, nitric acid, or other acids

**activation energy** input of energy required for a reaction to begin

**algal bloom** rapid increase in the number of algae. The bloom is often caused by increased nitrogen or phosphorus in the water.

**atom** smallest particle of a chemical element that still has the properties of that element. Atoms are considered the "building blocks" of matter.

**base** compound that accepts a hydrogen ion in a chemical reaction

**battery** two or more joined electrochemical cells

**bioluminescence** light released by certain chemical reactions in fireflies, some deep-ocean fish, and a few other living things

**buffer** compound that acts to keep pH constant when an acid or base is added to a solution

**calorie** amount of energy it takes to raise the temperature of one gram of water 1° C (1.8 ° F)

**carbon dioxide ($CO_2$)** gas in Earth's atmosphere. Its rising levels are cited as a cause of global warming.

**catalyst** material that speeds up a reaction by lowering its activation energy

**chemical bond** attraction between two atoms, formed by the sharing or transfer of electrons

**chemical change** change of matter's identity and properties, in which old chemical bonds are broken and new chemical bonds are made

**chemical equation** type of sentence that identifies the substances that enter into and are produced by a chemical reaction

**chemical reaction** *See* chemical change

**coefficient** number used in a balanced chemical equation to show the relative amount of a reactant or product

**combustion reaction** reaction in which a fuel combines with oxygen gas, forming carbon dioxide and water. It is also called a burning reaction.

**compound** when at least two different elements are joined in a chemical reaction, the molecule formed is called a compound

**conservation of energy** the principle that a chemical reaction neither creates nor destroys energy, but that energy changes form

**conservation of mass** the principle that a chemical reaction neither creates nor destroys mass. It is sometimes called the conservation of matter.

**decomposition reaction** reaction in which a single reactant decomposes (breaks apart) to form two or more products. It is the reverse of a synthesis reaction.

**double replacement reaction** reaction in which two reactants replace one another

**electrochemical cell** chamber in which redox reactions are used to make electricity

**electron** one of the very small, negatively charged particles that are a part of atoms, found outside the atom's nucleus

**element** pure substance that cannot be decomposed into simpler substances. Over 100 elements are organized in the periodic table.

**endothermic** describes a reaction that takes up heat

**energy** ability to do work. May take many forms, including light, heat, electricity, sound, or the potential energy of chemical bonds.

**enzyme** catalyst for the reactions of living things

**ethanol ($C_2H_6O$)** compound that can be made from corn or other plants. It is used as a fuel.

**exothermic** describes a reaction that releases heat

**formula equation** chemical equation written in chemical symbols, rather than words

**fossil fuel** coal, oil, natural gas, or other fuel made from the remains of living things

**fuel** any material burned for its energy

**global warming** gradual rise of average temperatures across Earth

**greenhouse effect** trapping of heat in Earth's atmosphere, caused by carbon dioxide and other gases

**Haber Process** process to increase yields of ammonia from a synthesis reaction. Named after its inventor, Fritz Haber.

**hydronium ion ($H_3O^+$)** ion found especially in acidic solutions

**hydroxide ion ($OH^-$)** ion found especially in basic solutions

**inorganic chemistry** the study of compounds not based on carbon

**ion** particle with an electric charge

**joule** unit of energy

**mass** amount of matter in something

**matter** the "stuff" that makes up everything in the universe

**methane ($CH_4$)** type of fossil fuel. It is also called natural gas.

**molecule** unit of matter, formed when chemical bonds join two or more atoms together

**nucleus** an atom's positively charged central core. It does not change during a chemical reaction.

**organic chemistry** the study of the carbon compounds of living things

**oxidation-reduction reaction** reaction in which one reactant is oxidized, meaning it loses electrons, and the other reactant is reduced, meaning it gains electrons. It is also called a redox reaction.

**oxygen** 1. an element 2. gas made of molecules of this element ($O_2$)

**periodic table** table that lists the known elements, organized according to each element's properties

**pH scale** scale used to measure the strength of an acid or base, or the acidity of a solution

**photosynthesis** process by which plants use energy from the sun to make food from raw materials

**physical change** change in the size, shape, or state of matter, but not in the identity of matter.

**plastic** type of human-made polymer, used in a variety of products

**polymer** large molecule made of repeated, small units joined together

**product** substance that a chemical reaction produces, or yields

**protein** essential nutrient that is needed for cell repair

**reactant** substance that enters into a chemical reaction

**redox reaction** *See* oxidation-reduction reaction

**respiration** process by which living things use oxygen to break down food for energy

**reversible reaction** chemical reaction that readily proceeds both forwards and backwards

**salt** class of compounds that includes sodium chloride. It may be a product of an acid-base reaction.

**single replacement reaction** reaction in which one reactant replaces another in a compound

**sodium chloride (NaCl)** table salt, one of many examples of a salt

**synthesis reaction** reaction in which two or more reactants join together to form a single product. This is the reverse of a decomposition reaction.

# Further information

## Books

*Chemical Reactions,* Louise Spilsbury (Heinemann Library, 2007)
*Chemicals in Action: Material Changes and Reactions*, Chris Oxlade (Heinemann Library, 2007)
*Fireworks!: Chemical Reactions*, Isabel Thomas (Raintree, 2007)

## Websites

**http://www.bbc.co.uk/schools/ks3bitesize/science/chemistry/chem_react_intro.shtml**
*This site provides a basic summary about chemical reactions.*
**http://www.chem4kids.com/files/react_intro.html**
*Learn even more about chemical reactions at this colourful site.*

**http://pbskids.org/zoom/activities/sci/**
*Mix up some bubbly chemical brews at this PBS website, while gaining some chemistry knowledge. The site includes other life and physical science projects you can do.*

**http://library.thinkquest.org/J001539/**
*Learn about the elements, do some chemistry experiments, and expand your knowledge of acids and bases on this website.*

## Place to visit

**Science Museum**
Exhibition Road
South Kensington
London SW7 2DD
www.sciencemuseum.org.uk
*Explore a museum dedicated to science!*

# Index